CATFISH FARMING

BARREND WILLEM LUDICK

authorHOUSE'

AuthorHouse™ UK
1663 Liberty Drive
Bloomington, IN 47403 USA
www.authorhouse.co.uk
Phone: UK TFN: 0800 0148641 (Toll Free inside the UK)
*   UK Local: 02036 956322 (+44 20 3695 6322 from outside the UK)*

Published by AuthorHouse 05/11/2021

ISBN: 978-1-6655-8920-8 (sc)
ISBN: 978-1-6655-8922-2 (hc)
ISBN: 978-1-6655-8921-5 (e)

Print information available on the last page.

This book is printed on acid-free paper.

Dedicated to the loving memory of

my son, Raymond Ludick

CONTENTS

PREFACE

I was born on the 26th of March 1942, on Carino Farm in Nelspruit, South Africa 26th March 1942. I was one of fourteen children. When I was 16, I went to work in Durban as an apprentice sheet-metalworker at Elgin Engineering. I married in 1967 in Durban. Then I began moving around and working in different provinces. I had four children. In 1980 I started working on the mines in Orkney in the North West province. During this time I became interested in fish breeding. I contacted Dr Ben van der Walt at the University of the North, as I read they had started experimenting with artificial insemination of the catfish (barbel). My wife and I were invited to attend a seminar on the breeding of catfish, and since that day, as the English would say, I was 'hooked'.

Through the coming years I experimented many times. Through trial and error—b few people were convinced that fish breeding was the future and could provide food for the masses. I dug dams at Orkney sewerage farm and started with duck before the fish project. Then the portion I was renting was reclaimed by the Orkney municipality for the enlargement of the sewerage plant, and financially I could not continue on my own without having land to farm on.

Since then I have had various ventures. I decided to carry out the breeding process on my own premises, as I enjoyed the wonder of artificial insemination breeding. I taught many people who were interested to see the process visually and practically. My son Raymond, was a keen participant in this venture. He enjoyed it and had his own set-up at his home. He passed away in a vehicle accident in 2004. I stopped actively breeding two years after his death.

I am 78 years of age and thought, why not write a book about my artificial insemination process skills? I have given a step-by-step guide, including photos taken with my late son. My wish is to inspire young entrepreneurs and hopefully show them that artificial insemination of breeding fish is not difficult.

INTRODUCTION

Why catfish farming,

The catfish is a tough and hardy fish that can withstand difficult environments. It has a respiratory organ and can survive in an atmospheric as well as an oxygenated system. A catfish will eat almost anything, so feeding is easy, but for quick growth, a high-protein diet is necessary. Catfish have good hearing, chemical and sensory antennae, with which they communicate.

The catfish farmer must look at the following factors when considering catfish farming:

- a constant water supply

- possible electricity failure (a generator to supply power in case of failure) condition of the ground, if outside breeding and growing are under consideration

- distance from marketing area, will involve transportation costs

- dangers presented by outside elements such as birds and otters, and the need for security

- a large building, for breeding and growing fingerlings before the fingerlings are transferred to outside dams; hygienic slaughtering premises; cooler rooms and other facilities for industrial storage

- weather conditions of your area

Breeding facilities are a number one priority to enable a constant supply of fingerlings. You can choose various types of breeding facilities for fingerling, such as dams, plastic containers, or steel frames with plastic linings. These containers are portable, leak-proof, and easy to clean and store. They are labour-efficient and cost-effective.

Fish farming falls under the Department of Agriculture, Forestry and Fishery's. Contact them for the necessary guidelines.

This breeding program is only for *Clarias gariepinus*, which survives in rivers, dams, reeds with water and sinkholes, and similar habitats.

The catfish species *Clarias gariepinus* is not found in the natural habitat of the Cape Rivers of South Africa. Transporting catfish to this area is prohibited by law.

When farming is considered in another country refer to your Department of Agriculture, Forestry and Fishery to determine if your area is suitable for this specific Species of Cat Fish. In general they thrive in warmer temperatures.

1. ARTIFICIAL BREEDING

It is wise to keep the males and females in separate portable brood stock containers for easy handling. The male has a small growth protruding, at the lower base of his body. It is easy to see. The female is smooth at the lower part of her body.

The ideal fish length ranges from 350 cm to 400 cm. The female can produce from 50,000 to 20,000 eggs at a time.

The best breeding time is September to November. Within seven days, the larvae resemble small fish.

Sort regularly and transfer to fingerling dams, and then to growing dams. They grow 200cm to 300 cm in the first year.

2. FROM EGG TO LARVAE

The biggest problem in South Africa is the demand for fingerlings. The process of artificial insemination has made obtaining fingerlings less difficult.

The female is treated to ripen the eggs, as explained fully in paragraph 9 and the male as well to ripen the sperm. The ideal temperature must be kept for the hatchery process. The breeding area must be properly equipped with all the necessary equipment, materials, and tools.

Hygiene is necessary to limit germs and infections. Wear rubber boots and protective clothing, and there must be no smoking in the area.

The breeding trays must be ready to receive the eggs for the hatching process. The trays are 300cm x 200cm x 50cm deep. The tray must have small holes in the bottom of the tray. Line the tray with a fine netting at the bottom and the sides. Stitch the netting by hand to the tray holes so that the netting does not lift up, when lowered into the breeding tank.

The female should be checked regularly once her abdomen starts to swell, with the ripening eggs. The eggs are later fertilized and put into the breeding trays to develop for 24 to 48 hours. This is fully explained once again in point no 9.

The Larvae will live off the egg sack for the first three days. Start feeding them food after this period every four hours. Within two weeks they should be about 25 to 30 cm long.

The catfish are cannibalistic, the bigger larvae tend to eat the smaller ones that is why it is important that they are sorted every two weeks, and put into fingerling dams when they are 40 to 50 mm long.

You can now market the fingerlings to future potential buyers or transfer them to larger dams for more growth.

3. DAMS AND CONTAINERS

Plastic dams, are easier in comparison to building cement dams, which can crack and leak as a result of the outside elements and weather. The costs are reduced considerably in this regard. You also have the element of nature—birds, hawks, otters, hail, theft—to contend with in outside dams.

The fish farm can be established in a large storeroom or building which supplies refuge from the elements for breeding and fingerlings and for growing dam containers.

Water temperature can be monitored more efficiently, and the fish are more visible.

The ideal depth for a dam is about one metre, so that the fish are clearly seen and are easy to remove with the least possible stress to them.

A generator is a good investment in case of power failures. This will ensure that electrical components for breeding will not be interrupted during the feeding and pump cycles.

4. FINGERLINGS TO HARVESTING

It is advisable to drain the dams every two months to sort the fish per size, or use a keepnet to remove and sort.

Cool days are ideal for this purpose. Cannibalism and growth rate can then be monitored, and feeding can be stepped up for quicker growth. Dirty water is an indication of too much food.

Take into consideration that it takes eight to ten months from breeding to reach a mass of plus minus 1 kg before you can harvest.

The fish grow slower during the winter months. They will grow quicker in a warmer controlled heated environment.

Fish harvested from mud dams should be put into freshwater dams and given no food for at least two days to remove the mud taste.

5. FEEDING GUIDELINES

The most expensive running cost will be catfish food. Pellets and other forms of food can be bought from your nearest Co-operations.

The larvae feed off the egg yolk sack for the first three to four days. They can then be fed fish flakes and tiny baby food pellets. When they are a length of 40 cm to 50 cm, they can be transferred to the fingerling tanks, where they can be fed larger pellets to sustain growth.

Catfish pellets are a very easy way to control the

feeding process. You can control and monitor the amount of food eaten and follow the growth process, to assess rate of growth in relation to the amount of food eaten.

A diet high in protein is best for fast growth of the fingerling to harvesting size.

Check regularly for infections and diseases and monitor ammoniac levels. Once the correct size is reached, transfer to growth dams.

Remember to sort fish according to growth size to prevent too much cannibalism.

6. MARKETING

The site of production must have a proper slaughtering facility. Hygiene is of the utmost importance. Safety rules and regulations must be strictly implemented. Refer to your Department of Land and Agriculture, Forestry and Fishery's for the specific rules and regulations as set out by each Country with regards to Fish Farming.

Bins and containers for all waste products must be on site. The waste products can be minced and used as a further method for feeding the larger fish. This is a good source of protein, except for the innards, they cannot be used at all. An industrial mincer will be ideal for this procedure.

Industrial cooler rooms are needed for preservation and storage of packaged products.

The easiest way to slaughter a catfish is to put the head into a vertical hook, cut around the base of the head, grip the skin with pliers, and pull towards the tail gently. Be careful not to rip the skin, as this will damage the flesh of the catfish.

There are various slaughtering procedures for the barbel:

- **Gutted.** Leave the head on, cut open the belly, and remove the innards.
- **Prepared.** Remove the head, skin, and innards.

- **Sliced.** Remove the head, skin, and innards, and cut in rings with a saw.

- **Filleted.** Remove the head, skin, and innards, and cut a long fillet along each side of the catfish's main backbone.

Package, and place in industrial freezers for marketing. Transport by means of refrigeration vehicles to deliver to the sources.

7. ILLNESS AND INFECTIONS

Catfish are hardy creatures and do not easily get sick. Because of intensive confined circumstances, stress is a factor.

Bacteria and viruses can be easily detected, and treated, when sorting is done.

They normally get white spots on them. Normally these parasites are brought in by fish eating birds, if farming outside.

The water must be checked regularly and also be

amoniac free, if it is as high as 7pp the fish will try to jump out of the tank.

This is an indication that something is not right.

A sudden lowering of temperature can cause great losses (± 15 - 16 °)

If you damaged the skin while handling, the skin will start peeling as infection sets in.

If you think it is an epidemic, get professional help immediately.

8. PORTABLE HATCHERY AND BREEDING OF CATFISH

The best time to breed catfish, is from September to November in South Africa.

Choose Male and Female brood stock, more or less the same size if possible. The best size is about 350 to 400 cm long.

These fish are easier to handle when using them for the artificial insemination process. Especially when injecting, slaughtering and the milking process.

Keep the males and females in separate portable brood containers, about 750 mm deep. They are easier to catch in the shallow water with a keep net, and are clearly visible.

Before injecting and treating the female or male, you must have the following ready, before commencing with treatment.

Portable brood tanks, connected, filled with water, the necessary pumps, filters, heaters, floating thermometer, and checked the PH and ammoniac levels of the water. The breeding trays in the tanks.

The bisecting table, should have the necessary tools, mutton cloth and waste containers nearby. All the necessary instruments and tools for the female and male, for injecting process, removing the testes, and then the removing the gland of the slaughtered male, and to preserve the pituitary gland for the next breeding period.

Knife, Plyers, scissors, tweezer, bowls, trays, injection

needles, surgical gloves, Slaughter trays, Saline water, spatula,carpet knife, hack-saw, testes grinder, bowl for testes, safety glasses, mutton cloth, towel and a small glass bottle with alcohol + - 100ml., all the necessary needed for the process.

Before commencing to inject the female, crush the pitituary gland and add 1cc of water from the tank, to the crushed gland and then suction it up into the needle holder.

The area where the fish is to be injected is shown clearly on the drawing of the fish from the side, and the picture below shows the markings from the halfway area of the fish to the tail area. This is where you will inject the mixture, then rub gently over the area, after injecting Stay away from the halfway area towards the head, you will puncture the lungs of the fish if you inject there. Once completed, return the fish to the brood stock portable dam.

1. Illustration from the side, inject between these markings.

2. Illustration from above, inject between these markings shown.

Make a note of the time you did this, so that the length of time in-between for injecting can be monitored. The ideal water temperature is from 25 to 28 degrees.

DAY 1

Inject the female at about 16h00 hours, with a crushed gland, mixed with 10cc water from the breeding tank. Refer to the illustrations of 1 and 2 to observe where to inject.

DAY 2

The next morning at 09H00 hours, inject the female, and the male with the same solution.

DAY 3

If the female's belly is swollen, the next morning, this is an indication that the eggs are ripening. Remove the female carefully from the brood-tank with the keep net. Cover her eyes with the mutton cloth. This calms her when handling. Put her onto a towel, dry the water from her body. The reason for this is: - that if the eggs get water in them while milking the female, the eggs will not be able to be fertilized, by the sperm. The eggs must be milked into a clean dry container. The colour of the eggs, should be a dark brownish colour if they are ripe, if not, she must be returned to the brood tanks for another hour, and then re-checked.

MILKING PROCESS

The male can then be removed from the brood tank with the keep net, cover his eyes with mutton cloth, to calm him.

Hold his body in a cloth, put him on the slaughtering tray.

Someone must assist you, as it cannot be done alone.

Make an incision at the lower belly with the carpet knife, take the scissors and cut it further open, the other person holding the catfish very firmly. Remove the testes and put into a bowl.

Take a clean sharp scissors and holding the testes over the milked eggs cut the testes and spread the sperm over the eggs.

Gently stir the eggs and sperm with a rubber spatula, for at least 60 to 80 seconds, to allow as many eggs as possible to be fertilized.

If the eggs are clustered together, after fertilization,

then add 1cc of saline water to the eggs and stir very gently, this is to loosen the jelly substance from the eggs.

The testes are ripe if they are light pink/orange in colour.

THE TESTES – NICE PINK COLOUR

Gently spoon the fertilized eggs directly into the prepared breeding trays, with circulating water until all the eggs have been added.

DAY 4/ 5

The hatching of the eggs is about 24 to 48 hours, depending on the temperature of the water. (25 to 28 degrees in the hatching dam)

The eggs must be covered with oxegenated water at all times while hatching.

You can remove an egg and put it under a microscope, or magnifying glass, to see if there is a visible red line.

This indicates that the egg has been fertilized, and is developing into a larvae.

DAY 6 /7

The larvae will start to move around after +_ 3 days, they will live off the yolk sack during this time. Then they will start to swim over the breeding trays into the tanks below.

The water should be at least 300 to 350 cm deep in the tank, with a lid to keep anything out which might put them in danger.

THE LARVAE

DAY 8/9/10

They can be fed Baby food for fish laying eggs, and egg powder. They will grow quickly in the next 20 days.

The fish should be sorted regularly, as they are cannibalistic and eat each other.

When they are about 20/50cm they can be transferred to the fingerling dams.

They must be sorted regularly as they grow, and transferred to dams to grow larger.

They can be harvested at a length of 350/400cm, for marketing purposes. A temperature of 16 degrees is very dangerous for the small fish.

REMOVING OF THE PITITUARY GLAND

Cut or saw off the head, just above the third eye called the FONTENELLE.

Just 10mm away from the side fins. Open the head structure, and remove the bone, until the brain is clearly visible. Lift the brain until you see the gland, it is about half the size of a match stick head.

The gland is connected to the brain with a fine visible thread. The gland is slightly lighter in colour than the brain.

Take an injection needle and lift the gland out, very gently so that you do not damage it. Put it into a glass bottle with alcohol to preserve it. It can be stored for up to a period of 1 year in the fridge. Write the date on the bottle.

9. SETTING UP A BREEDING HATCHERY

INTRODUCTION

One of the most common problems of fish farming, is the availability of fingerlings. This is where the process of breeding your fingerlings is important and to have a regular supply of fingerlings.

In 1987 the Americans market value for barbers was 168,000 tons of fish. Since then it improved vastly. The training of four persons, for this course will be for a period of ± two weeks, depending on the breeding process. This is from the beginning to completion, when eggs have hatched, and larvaes develop into small fish that can swim and start eating.

BEFORE YOU GET STARTED ON SETTING UP YOUR BREEDING HATCHERY

- 10 Cat-fish (Barbers), that have had the pitituary glands removed, and placed in a glass bottle in alcohol (80%) and placed in the fridge, do this before setting up the hatchery).
- 6 Cat-fish (Barbers) in small brood tanks, three males and three females, in separate tanks.

Make sure you understand the first two chapters before you buy any of the Materials and utensils.

ASSEMBLING OF THE NECESSARY HATCHERY BROOD TANKS

This is a suggested list, and can be adapted. The tank must be set-up and the water must be the correct heat and no amoniac in the water. All the materials and tools must be set out and easily available when needed. This list is for a farmer who has the capacity to farm on a large scale. For the person who wants to bread on a small scale, only buy one of each item (Small community Fish Farming).

YOU WILL NEED THE FOLLOWING MATERIALS AND UTENSILS TO PREPARE FOR BREEDING

Your own costing must be done on the specific list as show below:

Portable breeding tanks 650X325X305 cm

Canopy and led light

Portable stands

Heater 100W

Floating Thermometer

Small keep fish nets

Medium keep fish nets

Large keep net (for catching brood stock)

PH test kit

Bio Filters

Air pump 7500 double

Diaphragms 7500

Air stones

Airline Tubing

Non return valve

Baby fish food 10G

2 litre plastic bottles

Small glass bottles

Large spatulas

Milking bowls (for eggs)

Testes bowls

Large Cutting knives

Kitchen scissors (meat cutting)

Spice crushing bowls with grinder

(For glands)

Tweezers

Containers to crush gland

(Small spice crushes)

Plyers

Small hacksaws

Carpet knives with extra blades

Rolls of mutton cloth

Towels (for fish)

Breeding trays + netting

1 X large long bisecting table (optional)

R100.00 Large tray for removing glands and testes

Bar Freezer

(If glands are removed immediately, not necessary)

Box of injection needles, plus holders

Alcohol

Salt

Larvae food

Box of Disposable injection holders

Box of 2.1 gauge injection needles

Box of disposable Gloves

Double adapters -two and three way.

(Used for pump, heater etc.)

Refuse bags for waste

Bucket with lids

(For waste products when slaughtering)

Safety glasses (clear)

Labels

(For dates of glands harvested)

A small solar heating system could be set up, for the breeding process, where normal electricity is not available.

Printed in the United States
by Baker & Taylor Publisher Services